W9-CKP-616

SPIDERS

Design
David West
Children's Book Design
Illustrations
Aziz Khan
Picture Research
Cecilia Weston-Baker
Editor
Denny Robson
Consultant
John Stidworthy

Designed and produced by
Aladdin Books Ltd
70 Old Compton Street
London W1

First published in the United States in 1988 by
Gloucester Press
387 Park Avenue South
New York, NY 10016

Printed in Belgium

ISBN 0-531-17077-2
Library of Congress Catalog
Card Number: 87-82900

This book is about all the different types of spiders in the world. It tells you what they look like, what they eat, and where and how they live. Find out some surprising facts in the boxes on each page. The Identification Chart at the back of the book will show you several common and some exotic species. It also indicates where they are to be found.

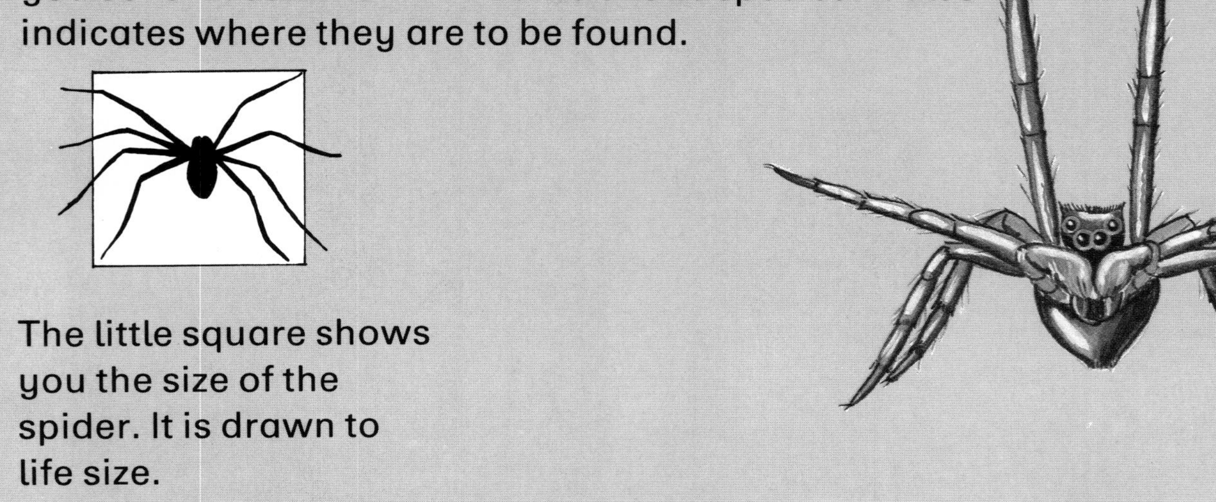

The little square shows you the size of the spider. It is drawn to life size.

The picture opposite shows a jumping spider

FIRST SIGHT

SPIDERS

Lionel Bender

GLOUCESTER PRESS
New York · London · Toronto · Sydney

Introduction

There are at least 50,000 different kinds of spiders in the world. They may live for up to 20 years and are found in every habitat and on all continents except Antarctica.

Spiders are all predators – they catch and eat other animals. Some, such as tarantulas, occasionally prey on mice and birds. Spiders are also likely to eat each other. But most spiders feed on insects. They eat these in such large numbers that they help to control insect pests. A square yard of grassland may contain more than 500 spiders which must eat hundreds of thousands of insects a year. Spiders usually catch their prey by using the silk they produce. They use silk in a variety of ways – to make webs, as a safety line when climbing and as a means of air travel.

Contents

◁ **A Lichen Wolf Spider**

Life with eight legs

Spiders are not insects. They always have eight jointed legs, not six as insects have, and they never have wings. The sensory organs on their head are not antennae but leg-like structures called palps. Spiders all have a pair of poison fangs and several pairs of spinnerets which produce silk.

A spider's body is in two main parts. The front part is a combined head and chest. It is joined to the back part, the abdomen, by a narrow waist. The mouth and jaws are on the underside of the head and the eyes are usually on the top. The legs are attached to the thorax, the chest, which contains the stomach. The abdomen contains the other major body organs.

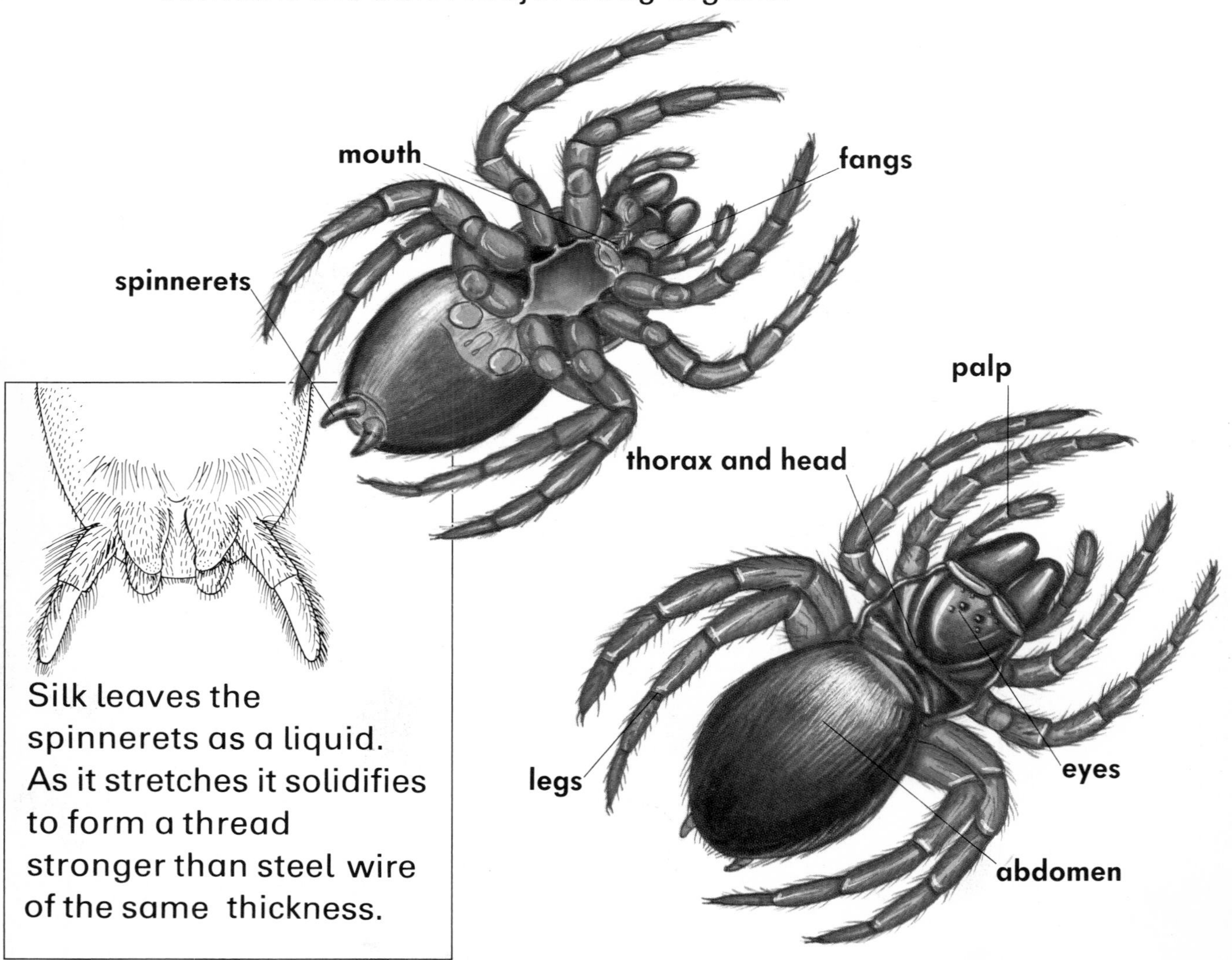

Silk leaves the spinnerets as a liquid. As it stretches it solidifies to form a thread stronger than steel wire of the same thickness.

Making orb webs is just one of the ways spiders use their silk ▷

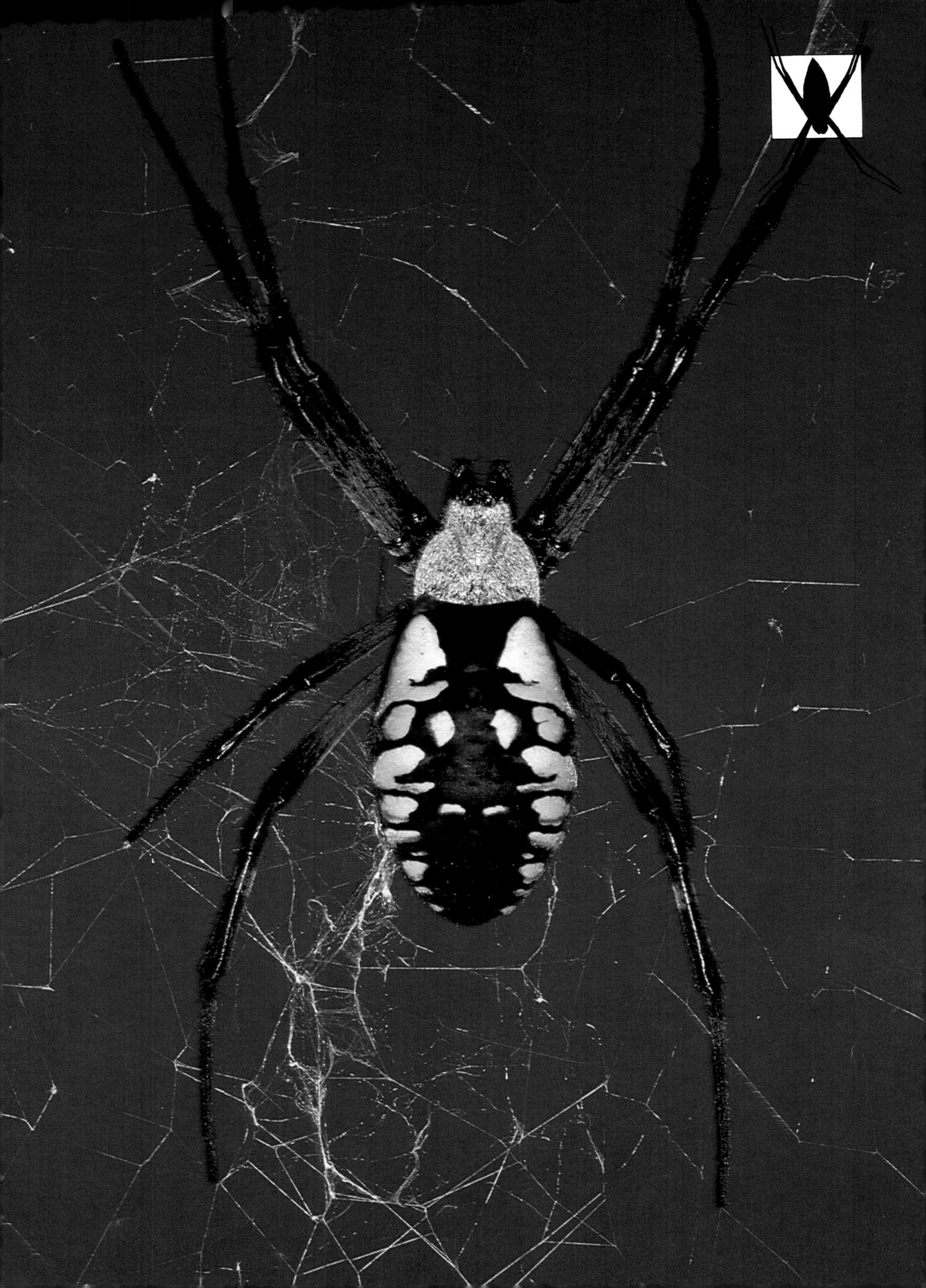

Fangs and jaws

Nearly all spiders use poison to stun or kill their prey. Bird-eating spiders and their relatives have very large, downward-pointing fangs that they raise and then thrust into prey. But generally spiders have a pair of small fangs that are used like pincers, biting inward toward one another.

Spiders can take food only in liquid form. Some, like crab spiders, inject digestive juices into prey through the hole left by their fangs. They then suck out the digested food, leaving an empty husk. Others chew prey with the tooth-like projections on their jaws. As they do this, they pour out digestive juices over the prey then suck up the meal from its mushy remains.

Each jaw has a poison gland from which a duct runs to the fang tip. A spider uses its fangs like a doctor's hypodermic needle to inject poison into the prey. Near the base of the fang there may be a set of "teeth."

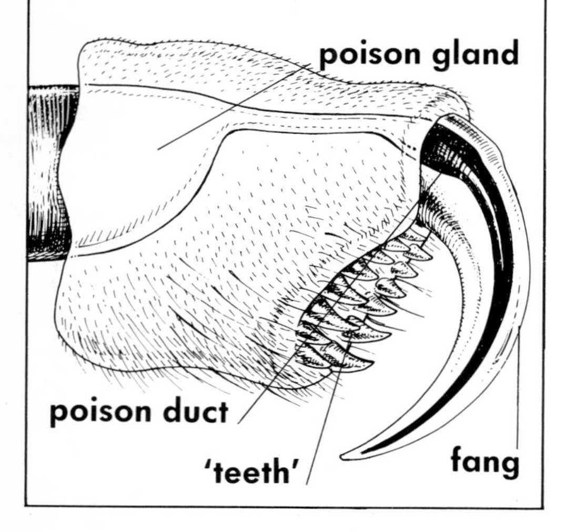

Bird-eating spiders have downward pointing fangs

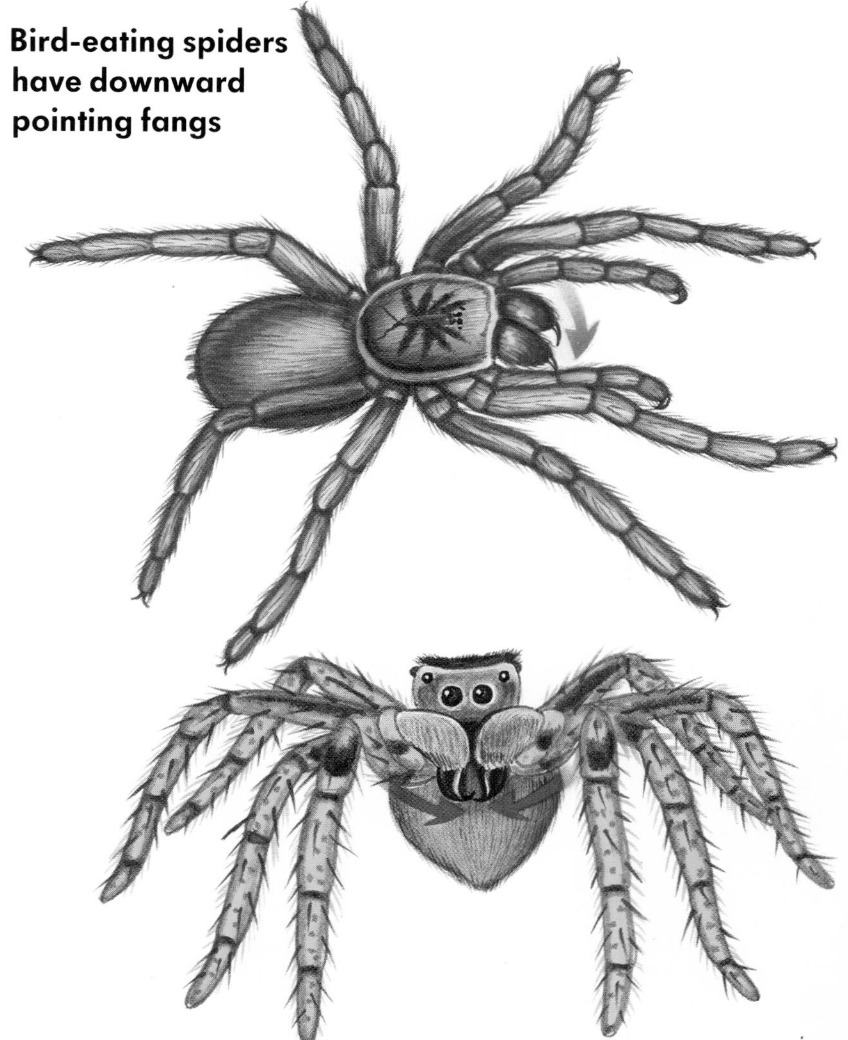

Jumping spiders have fangs that work sideways

A male Green Orb Spider courts a female by vibrating the web ▷

Male and female

Male spiders usually have smaller bodies than females, although their legs may be as long. The female Common Garden Spider, for example, is 1½ times the length of her mate and much more bulky. In some species the difference in size may be so great that the males are possible meals for the females.

Before mating, a male spider makes a small silk web on which he deposits sperm. He dips his palps into the sperm and then finds a female. He signals to her so that she recognizes him as a suitable partner and he persuades her not to eat him. During mating, he transfers the sperm from his palps to her reproductive opening. Then she may eat him.

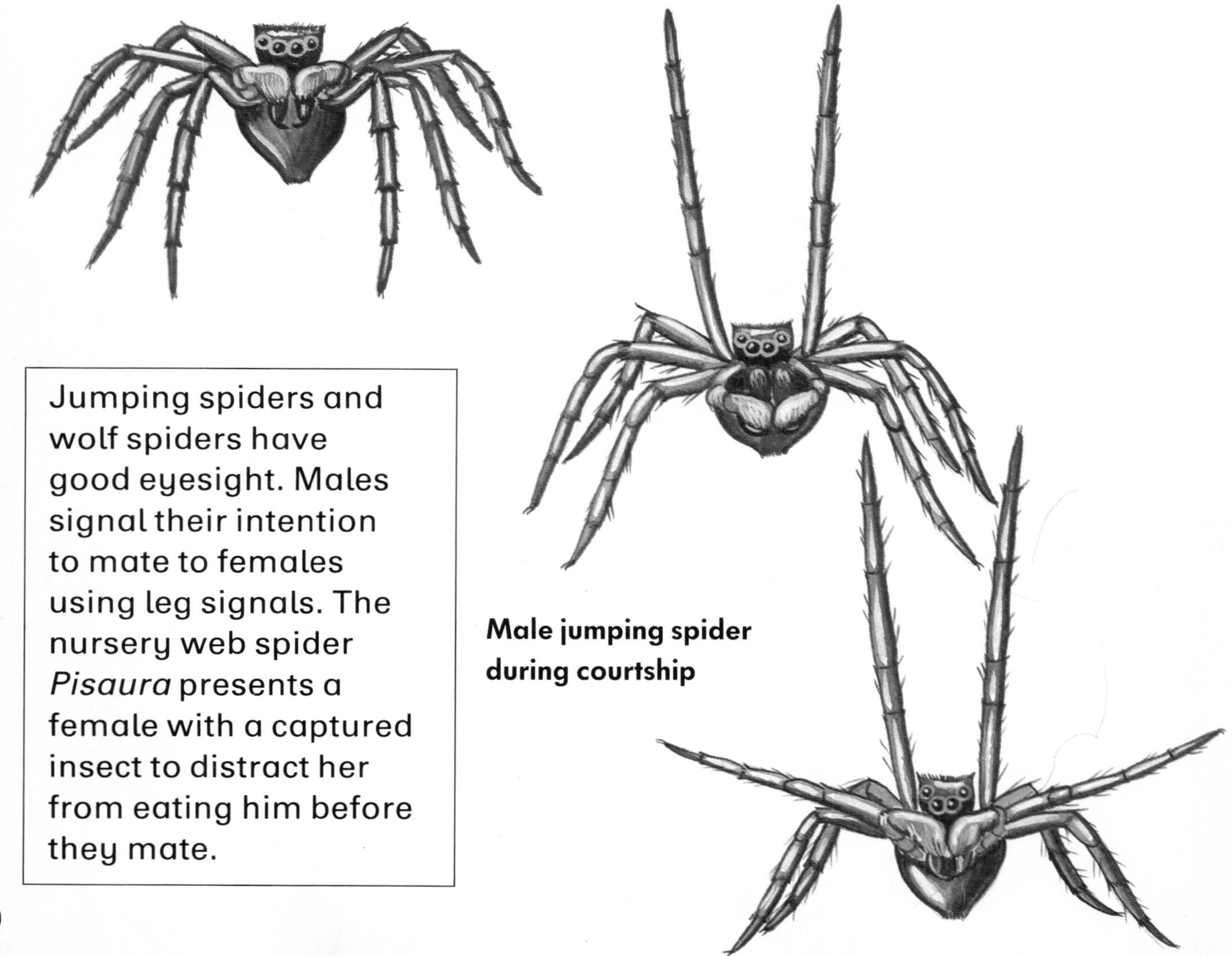

Jumping spiders and wolf spiders have good eyesight. Males signal their intention to mate to females using leg signals. The nursery web spider *Pisaura* presents a female with a captured insect to distract her from eating him before they mate.

Male jumping spider during courtship

Eggs and young

The little spider *Oonops* lays just two eggs. Others, like some of the Orb-web spiders, lay more than 1,000. The eggs are usually laid on a plant. Mother spiders guard the eggs or wrap them in a protective web or in a silken cocoon. Many, like the female wolf spiders, carry the cocoon around with them.

At hatching, baby spiders do not have spinnerets or poison and they cannot see very well. They survive on their egg yolk until they shed their cuticle (the hard outer skeleton) and become tiny versions of their parents. They leave the nest by climbing to the top of the plant and letting out a silk thread. On this they are blown by the wind to a new home.

To grow, a spider sheds its cuticle and rapidly expands before the new one hardens. It may go through ten such molts.

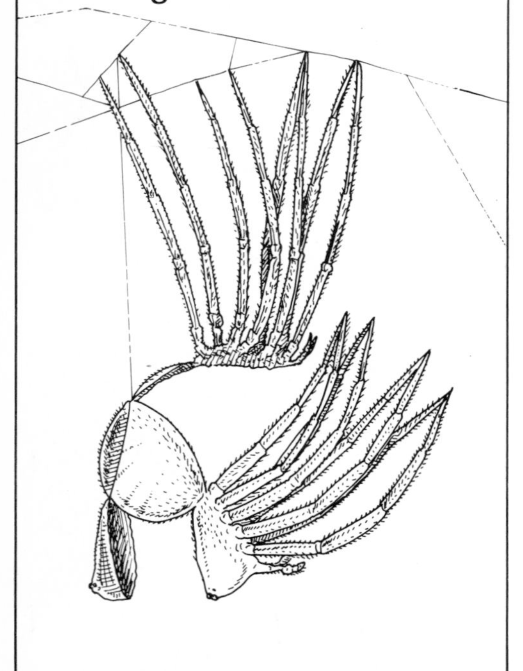

Dolomedes with its cocoon of eggs

A nursery full of baby Garden Spiders ▷

◁ **The spines and shape of *Gasteracantha* give it protection from attackers**

Defense

Many animals prey on spiders. Birds and shrews, for example, are not put off by the spider's poisonous bite. A spider's enemies also include larger spiders and some insects, like hunting wasps. But spiders have developed various means of defense.

Crab spiders and most garden spiders are colored and patterned to blend in with their background. Other spiders have odd shapes which give them protection by confusing their predators. A rare trapdoor spider has two lines of defense – a trapdoor to its burrow and a flat, corky abdomen. When the spider is touched, it runs down into its burrow and presents its 'shield' to the attacker.

Bird-eating spider

The large bird-eating spiders of the tropics have a poisonous bite. But they often deter attackers by using their legs to kick off hairs from their backs. These hairs can irritate eyes, nose and skin.

This trapdoor spider has a hard flat abdomen to protect it from its enemies

Bird-eating spiders eat the largest prey. They mainly eat small birds. But some have been known to catch frogs, lizards and even snakes.

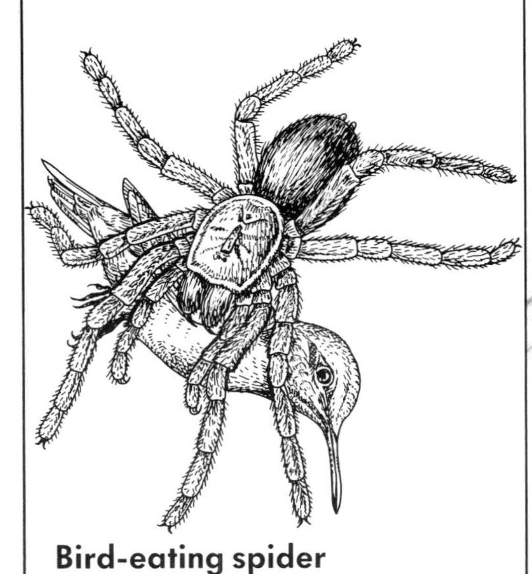

Bird-eating spider

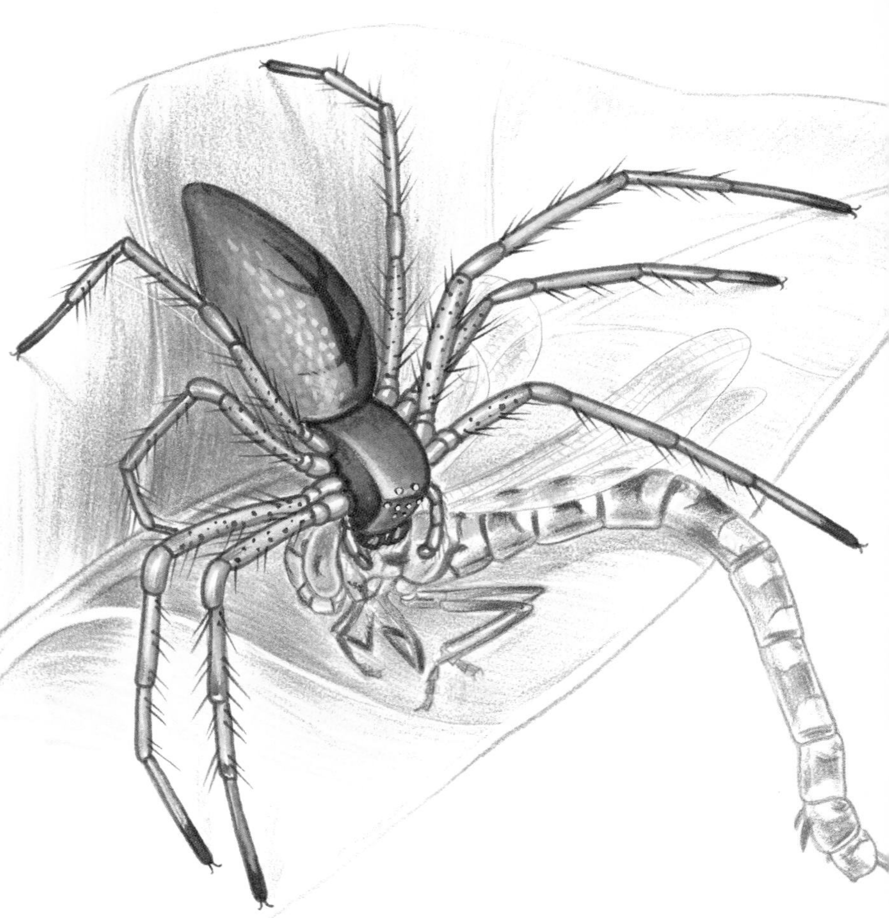

A lynx spider overpowers a dragonfly

Pursuing prey

Wolf spiders are fast runners. They can chase and capture prey, including large and fierce beetles. Jumping spiders also hunt down prey rather than wait for it to be tangled in a silk web. They creep up on a possible victim and then pounce and grab it with their front legs. Jumping spiders have extremely good eyesight and can judge distances well. They rarely miss their target.

Lynx spiders use their long legs to run after or leap on their prey. Other spiders simply lie in wait in their lairs and then pounce as their prey passes by. The Water Spider, common in ponds and lakes, and some of the large bird-eating spiders catch prey in this way.

A Water Spider takes a captured worm to its air-bubble lair ▷

A robber fly fails to notice a crab spider and is captured ▷

Spider tricks

As well as being used for protection, camouflage can be used as a weapon to help catch prey. The dull colors of many garden spiders, and their stillness, allow them to trap unwary insects. Brightly colored spiders can also be invisible to their prey. Sitting on flowers of the same color, crab spiders simply wait for insects to walk into their arms.

Some grass spiders resemble twigs, sticks and grass and go unnoticed until it is too late for their victims. A few tropical forest spiders look like foul-tasting insects, like ants, which deters predators. The *Myrmecium* spider looks so much like the ants on which it feeds that it can live in the ant nest.

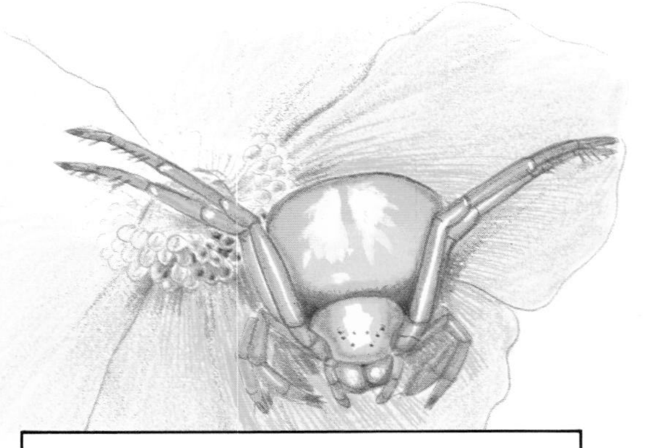

Crab spiders are adapted to the flowers they live on

Ant-mimic Spiders wave their front pair of legs about to look like an insect's antennae. They fool their victims and so can get close enough to attack them.

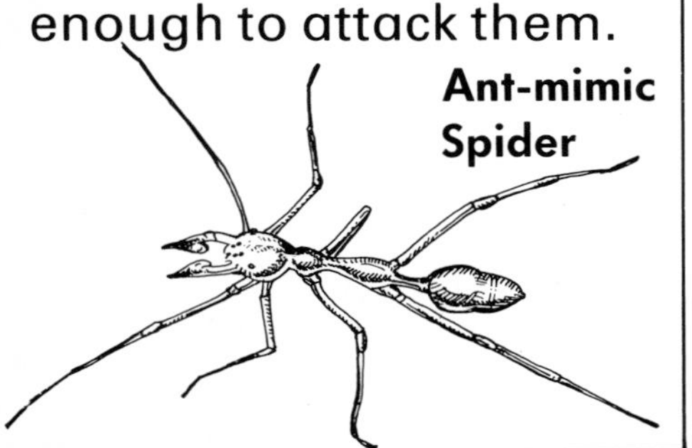

Trapdoors and purses

Trapdoor spiders live mainly in tropical regions. Each digs a burrow, caps it with a hinged trapdoor of silk and then lies in wait out of sight. When the spider detects the vibrations of passing prey, it flips up the door, pounces, then pulls its victim into the burrow and eats it. Trapdoor spiders are most active at night.

Purse-web spiders usually live in warm regions. They line their burrows with silk and extend the silk over the surface of the ground to make a pouch or purse. When insects stumble over the pouch, the spider comes out from the burrow and kills its prey from inside the pouch. It then pulls its victim through the silk and goes back underground to feed.

A purse-web spider traps its prey

Purse-web spiders use their huge fangs in many ways – to dig out a burrow, gently place earth on the pouch for camouflage, and to stab, poison and hold their victims. They cut a hole in the pouch using their "teeth."

An Argentinian trapdoor spider grips its door in anticipation ▷

Many money spiders do not need sticky thread to catch prey. They sit upside down below a silk "hammock." Insects blunder into threads above this and fall to their doom.

A comb-footed spider waits in its tangled web ▷

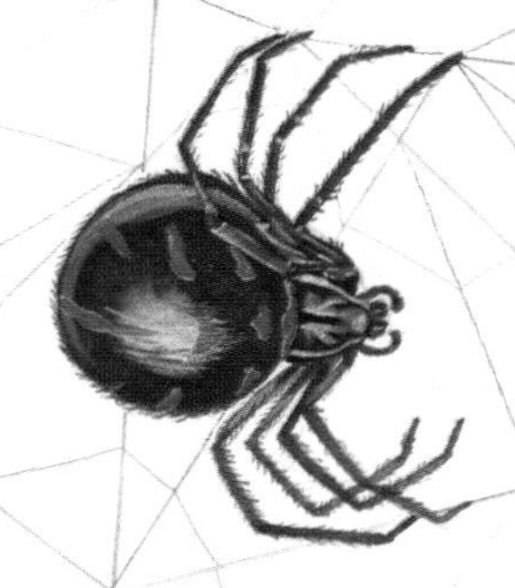

Black Widow Spider

Scaffolds and hammocks

The comb-footed spiders, which include the dangerous Black Widow, make webs that look like untidy scaffolding. There is usually a central maze of threads where the spider lurks and various sticky threads on which victims get trapped.

In some of these scaffold webs flying insects are caught by outer sticky threads. In others, the middle of the web is thickened and long threads hang down to the ground. At the ends of these the spider places sticky drops. As a crawling insect blunders into a thread it snaps. Being elastic the thread shortens and the victim is pulled helplessly off its feet. The spider then hauls up its catch and eats it.

Orbs

The orb web is made by many species. It looks like a bull's-eye and is the web most associated with spiders. Spiders weave the orb in simple stages. A young spider has this ability from hatching.

First the spider makes a bridge between two supports. It often relies on wind to carry a thread across. Then it makes a Y-shape that will be the center of the web, the hub. From this it attaches other threads to the supports. It moves outward from the hub to the edges, making the spiral stronger. Finally, it spirals inward, laying down the sticky silk that traps insects. Many species hide near the web keeping hold of a signal thread with their legs.

An orb spider constructs its web

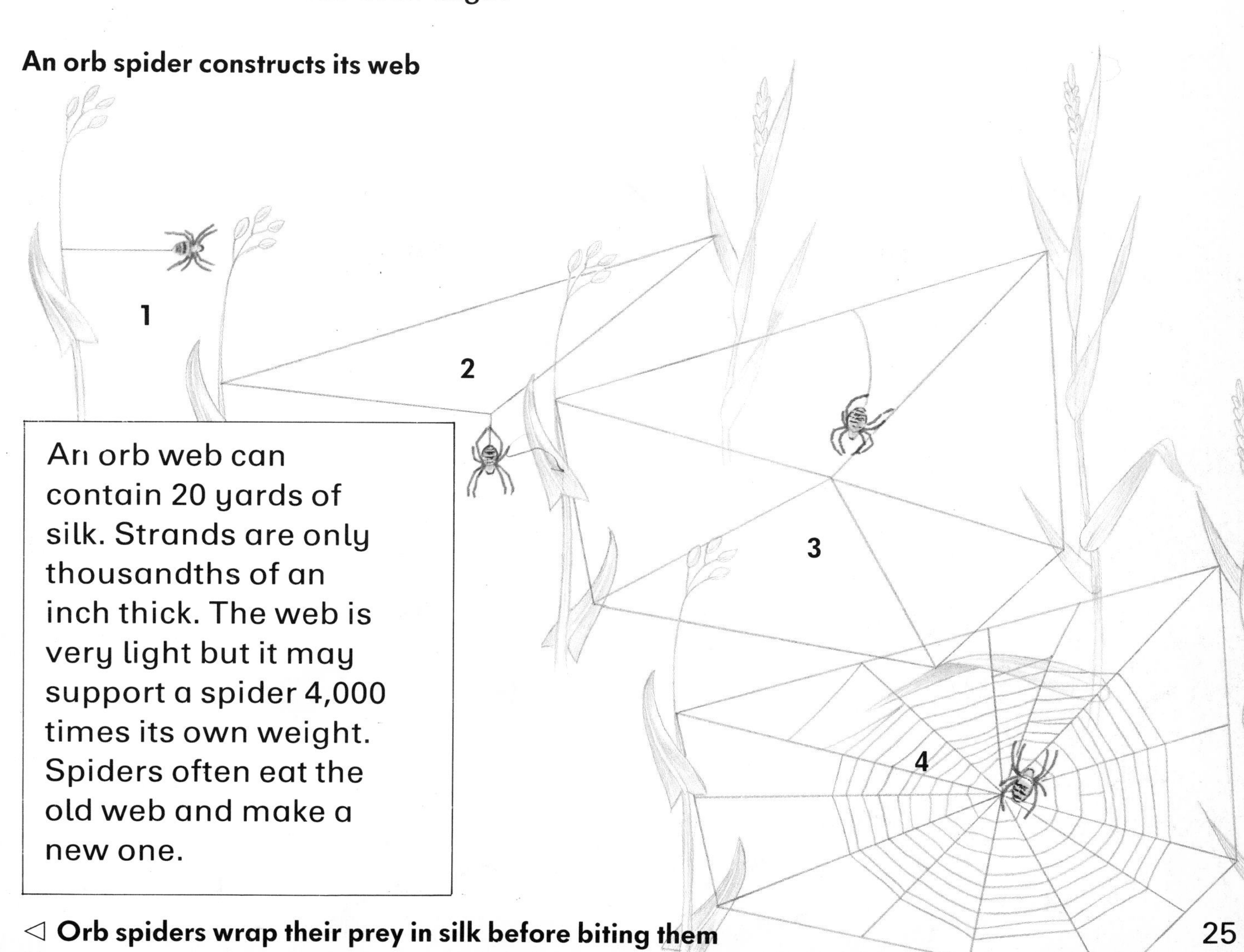

An orb web can contain 20 yards of silk. Strands are only thousandths of an inch thick. The web is very light but it may support a spider 4,000 times its own weight. Spiders often eat the old web and make a new one.

◁ **Orb spiders wrap their prey in silk before biting them**

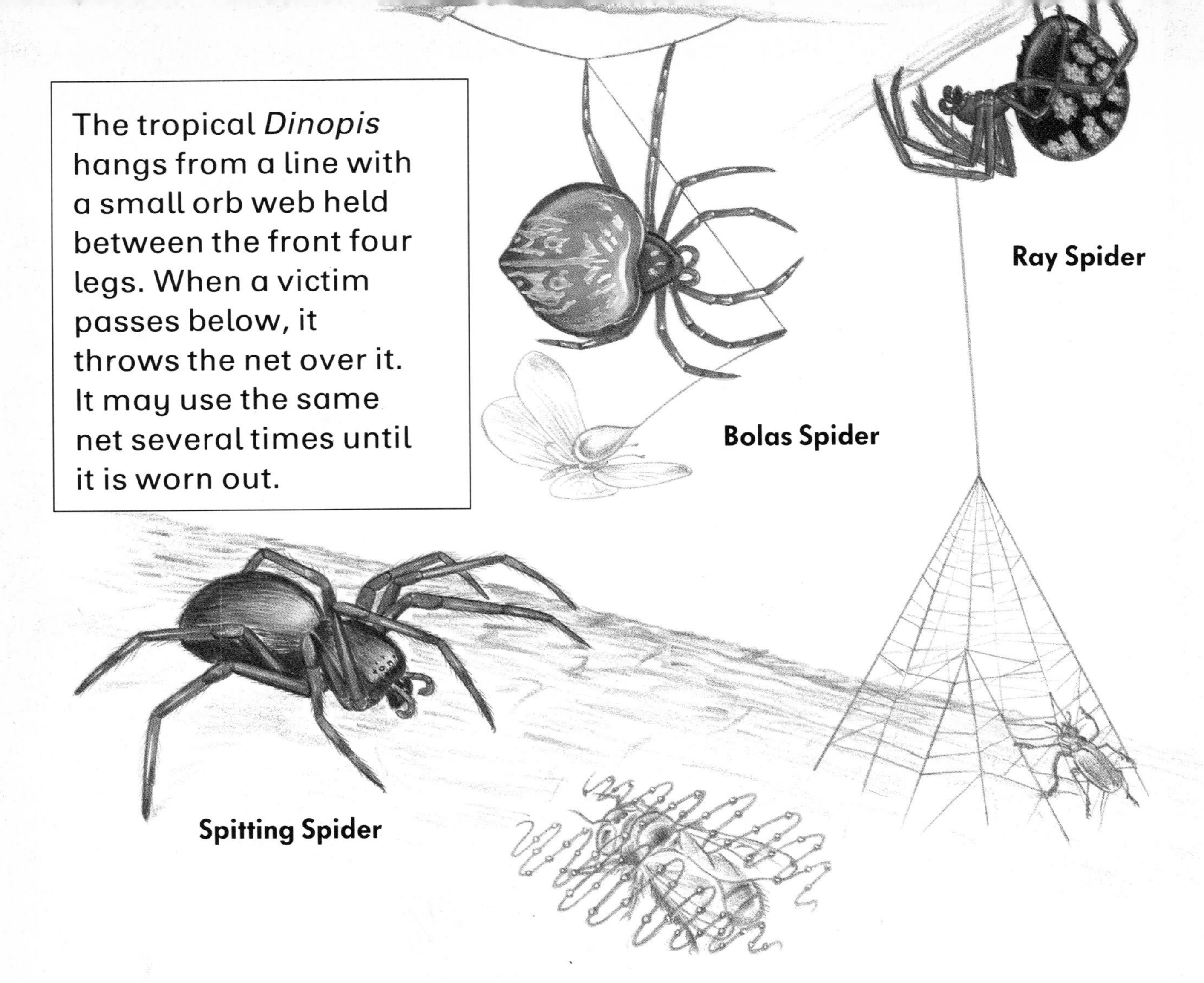

The tropical *Dinopis* hangs from a line with a small orb web held between the front four legs. When a victim passes below, it throws the net over it. It may use the same net several times until it is worn out.

Capture at a distance

Most web-making spiders wait for prey to become entangled in their snares. But some take a more active role, like the Bolas Spider of the tropics. This spider puts a blob of sticky silk on the end of a line hanging down from its web. It twirls this thread around until it catches a passing insect.

The spitting spider *Scytodes* has an even stranger way of catching prey. It is only ·2in (5mm) long and moves very slowly, yet it can catch fast-moving prey much larger than itself. It sprays them with a mixture of poison and gum from its fangs, over a distance of ·4in (1cm). This pins down the insect. The spider then walks up to its victim and kills it with its bite.

A net-casting spider (*Dinopis*) with its web waiting for a victim ▷

Survival file

Some people dislike spiders. Others are frightened by them, even though only a few spiders are actually dangerous to people. Some spiders are thought to be bringers of good fortune and killing spiders is thought to bring bad luck. Many people think it wrong to kill a spider because of the good they do in getting rid of insect pests. However, few people would think of spiders as endangered species. Yet many species are rare and in danger of extinction. When forests are cut down or grassland is destroyed, some of the best habitats for spiders disappear.

A Funnel-web Spider with its lizard prey

Another danger for spiders is the insecticides people spray on crops or gardens to kill pests. Spiders are killed directly by the chemicals and others die because of the loss of their insect food. But other spiders thrive living close to people. In northern Europe, the spitting spider *Scytodes* and the long-legged *Pholcus* live only in houses. The large hairy house spiders also benefit from their association with people.

Some spiders are not welcome near people. The Black Widow Spider of warm temperate climates is feared because of its poisonous bite. It is one of the few species that can be fatal to humans. Its habit of making its home in garages makes this species a problem to people living in parts of North America and Europe.

Another dangerous spider is the Funnel-web of Australia. This is a relative of the bird-eating spiders and is sometimes found in gardens. It is more aggressive than most spiders. The bird-eating spiders themselves can be as much as 8in (20cm) across the legs. But despite their appearance they are rarely aggressive. And although their bite is powerful, it is rarely dangerous for a human.

A female Black Widow Spider

Insecticides kill spiders as well as their prey

Identification chart

Some of the main types of spiders are shown here drawn to scale. The grid is divided into 7mm squares. Large and spectacular kinds like the Golden Silk Spider and Jumping Spider may be seen in a zoo. You can see others like Water Spiders and Orb Web Spiders in your gardens and parks. Be careful not to destroy their habitat.

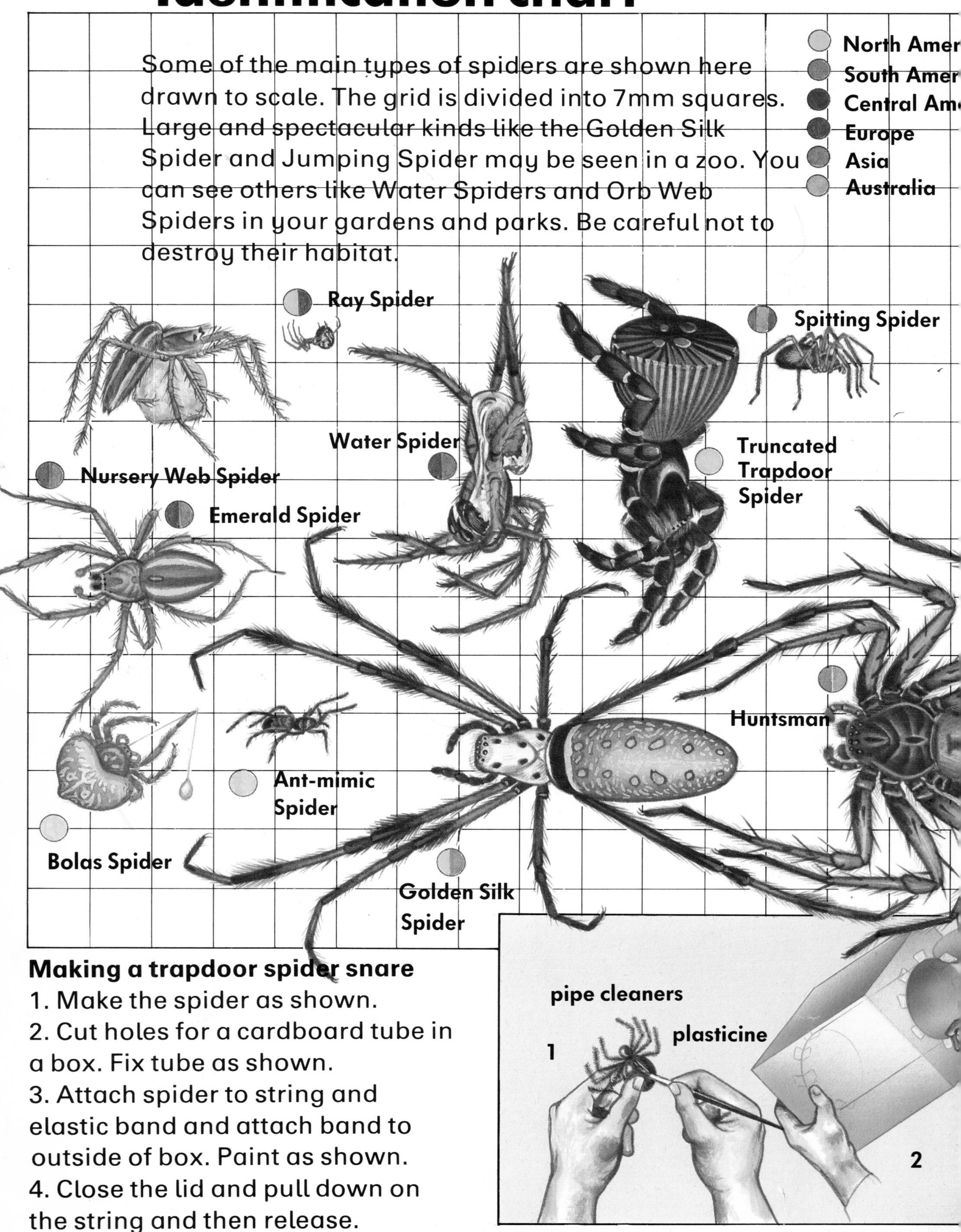

Making a trapdoor spider snare

1. Make the spider as shown.
2. Cut holes for a cardboard tube in a box. Fix tube as shown.
3. Attach spider to string and elastic band and attach band to outside of box. Paint as shown.
4. Close the lid and pull down on the string and then release.

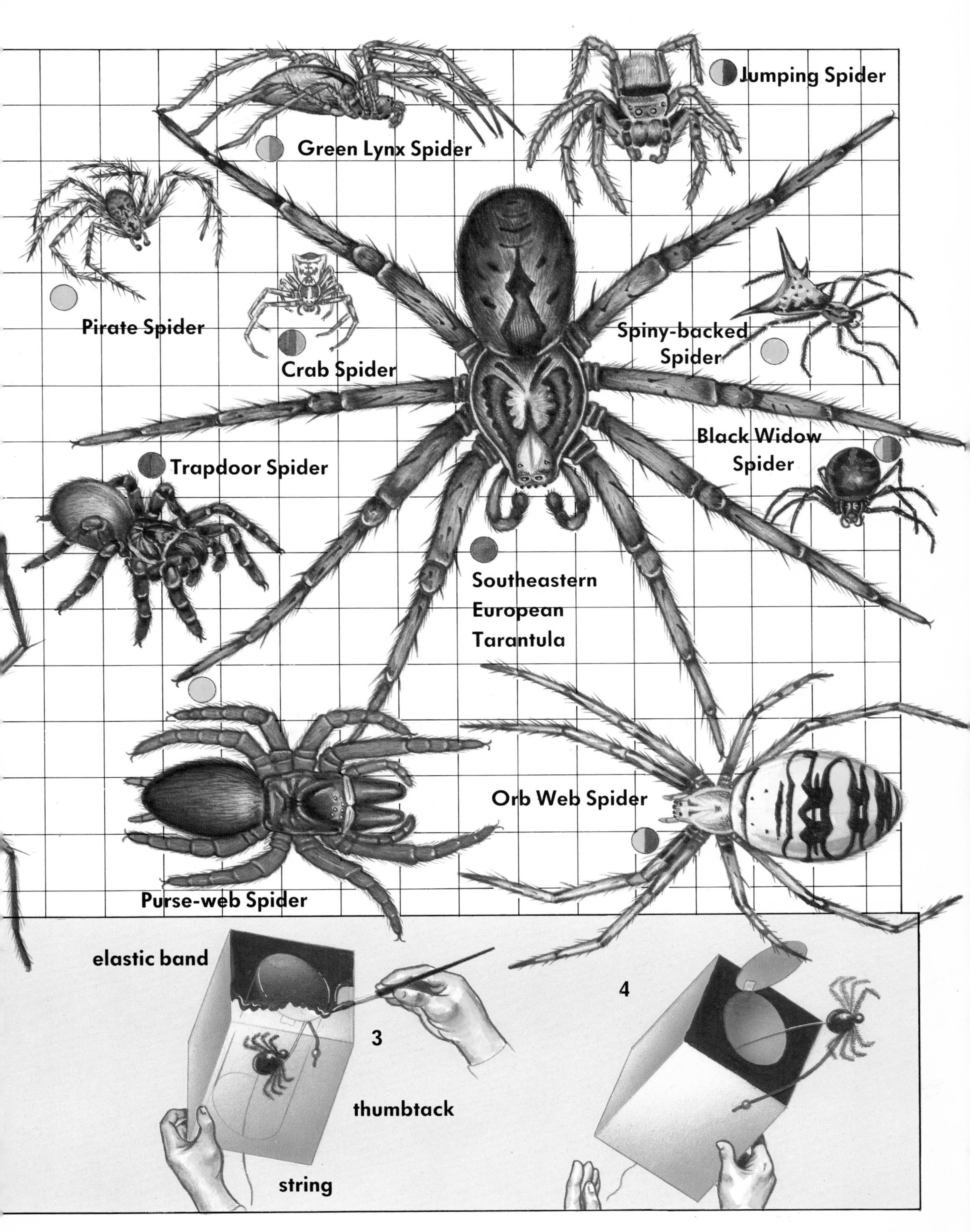
Jumping Spider
Green Lynx Spider
Pirate Spider
Crab Spider
Spiny-backed Spider
Black Widow Spider
Trapdoor Spider
Southeastern European Tarantula
Orb Web Spider
Purse-web Spider
elastic band
3
4
thumbtack
string

Index

The picture on the cover shows the spider *Argiope argentata*

Photographic Credits:
Cover and pages 7, 9, 11, 17, 21, 23, 24, 27, 28 and inset: Bruce Colman; title page: Survival Anglia; intro and page 13: Ardea; pages 14 and 29: Planet Earth; page 19: Robert Harding.

PRINTED IN BELGIUM BY proost INTERNATIONAL BOOK PRODUCTION